NOTES TO USERS

THESE Notes are complementary to A.P. 2095
Pilot's Notes General and assume a thorough
knowledge of its contents. All pilots should
be in possession of a copy of A.P. 2095 (see
A.M.O. A718/48).

Additional copies may be obtained by the
station publications officer by application on
Form 294A, in duplicate, to Command head-
quarters for onward transmission to A.P.F.S.
(see A.P.113.) The number of this publication
must be quoted in full—A.P. 2062, A, C, F,
H, K, L & M.—P.N.

Comments and suggestions should be for-
warded through the usual channels to the Air
Ministry (T.F.2.).

AIR MINISTRY   A.P. 2062, A, C, F, H, K, L & M—P.N.
March, 1949   Pilot's Notes
(4th Edition)
(Reprinted—October, 1949)

LANCASTER Mks. 1, 3, 7 and 10
PILOT'S NOTES

LIST OF CONTENTS

PART IV—EMERGENCIES

PART V—ILLUSTRATIONS AND LOCATION OF CONTROLS

LANCASTER
(ALL MARKS)
PILOT'S CHECK LIST

(Excluding Checks of Operational Equipment)

ITEM	CHECK
1. Weight and balance.	Within permissible limits.
2. Authorisation book.	Sign.
3. Form 700.	Sign.
External checks.	
N.B.—Start at the entrance door and work clockwise around aircraft.	
4. Dinghy external release.	Secure.
5. Tail wheel.	Extension of oleo leg. Tyre for cuts and creep. Valve free.
6. Starboard tailplane.	Condition of upper and under surfaces. Leading edge.
7. Starboard fin.	Condition. Leading edge.
8. Starboard rudder.	Condition. Trimmer. External control lock removed.
9. Starboard elevator.	Condition. Trimmer. External control lock removed.
10. Rear turret.	Locked fore and aft.
11. Rear lights.	Condition.
12. External aerials.	Condition.
13. Port elevator.	Condition. Trimmer. External control lock removed.

ITEM	CHECK
14. Port rudder.	Condition. Trimmer. External control lock removed.
15. Port fin.	Condition. Leading edge.
16. Port tailplane.	Condition of upper and under surfaces. Leading edge.
17. Static vent.	Plug removed.
18. Port mainplane.	Condition of upper surface. Fuel tank filler covers secure.
19. Port flaps.	Position. Condition.
20. Port aileron.	Condition. Trimmer. External control lock removed.
21. Port navigation light.	Condition.
22. Port identification lights.	Condition.
23. Port mainplane.	Condition of leading edge. Condition of undersurface.
24. Landing lamps.	Condition. Position.
25. No. 1 engine.	Security of oil tank and coolant filler covers. Security of cowlings. Condition of propeller and spinner. Oil and coolant leaks.

ITEM		CHECK	ITEM		CHECK
26.	No. 2 engine.	Security of oil tank and coolant filler covers. Security of cowlings. Condition of propeller and spinner. Oil and coolant leaks.	36.	Starboard fuel jettison pipe cover plate.	Pin secure.
27.	Port undercarriage.	External lock removed. Microswitches clean and free. Extension of oleo legs. Brake leads secure. Towing shackle secure. Locking ring for creep and wedges in position. Valve free. Tyre for cuts and creep. Chock in position.	37.	Starboard undercarriage.	External lock removed. Microswitches clean and free. Extension of oleo legs. Brake leads secure. Towing shackle secure. Locking ring for creep and wedges in position. Valve free. Tyre for cuts and creep. Chock in position.
28.	Port centre section.	Leading edge panel secure. All screws flush fitting.	38.	No. 3 engine.	Security of oil tank and filler covers. Security of cowlings. Condition of propeller and spinner. Oil and coolant leaks.
29.	Port fuel jettison pipe cover plate.	Pin secure.	39.	No. 4 engine.	Security of oil tank and filler covers. Security of cowlings. Condition of propeller and spinner. Oil and coolant leaks.
30.	Trailing aerial fair lead.	Condition.			
31.	Bomb doors.	Condition.			
32.	Pressure head.	Cover removed.	40.	Starboard mainplane.	Condition of leading edge. Condition of undersurface.
33.	Nose.	Condition.			
34.	External fire-extinguishers.	In position.	41.	Starboard navigation light.	Condition.
35.	Starboard centre section.	Leading edge panel secure. All screws flush fitting.	42.	Starboard identification lights.	Condition.

	ITEM	CHECK		ITEM	CHECK
43.	Starboard aileron.	Condition. Trimmer. External control lock removed.	60.	Crash axe.	In position.
			61.	Emergency pack.	Stowed.
44.	Starboard flap.	Position. Condition.	62.	Dinghy radio.	Stowed.
45.	Starboard mainplane.	Condition of upper surface. Fuel tank filler covers secure.	63.	Main oxygen supply.	On.
			64.	Nitrogen main cock (if fitted).	On if required.
46.	Static vent.	Plug removed.			
47.	Dispersal area.	All clear around aircraft.	65.	Mid escape hatch.	Operation. Secure.
	Internal checks.		66.	Hydraulic accumulator.	Pressure 220 lb./ sq. in.
	N.B.—Start at the rear of the aircraft and work forward.		67.	Emergency air system.	Pressure (if gauge fitted) 1,200 lb./sq. in. approx.
48.	Rear turret.	Doors closed. Position of "dead man's" handle.			
49.	Fire-extinguisher.	In position.	68.	Ground/ flight switch.	Flight.
50.	Elsan.	Secure.	69.	Electrical panel.	All circuit breaker switches in (Mk. 10 only). Battery state. Generator warning lights on. Generator switches on. Earth lights.
51.	Dipsticks.	In position.			
52.	First-aid kit.	In position.			
53.	Flying control rods and trimmer cables.	Free from obstructions.			
54.	D.R. compass master unit.	Clear of magnetic interferences.	70.	Batteries.	Leads secure.
			71.	Intercom.	Switch on.
55.	Loose equipment.	All secured.	72.	Crossfeed cock.	Freedom of movement. Turn off.
56.	Crash axe.	In position.			
57.	Fire-extinguisher.	In position.	73.	Fire-extinguisher.	In position.
58.	Rear escape hatch (if fitted).	Operation. Secure.	74.	Fuel gauges.	Switch on (Mks. 1, 3 and 7). Contents.
59.	Mid upper turret.	Condition.	75.	Booster pumps.	Test operation by ammeter and switch off.

8

ITEM	CHECK
76. Pressure-head heater.	Test operation by ammeter. Switch off.
77. Tank selector cocks.	As required.
78. Fuel flowmeters (if fitted).	Set to zero.
79. Emergency air selector knob.	Off.
80. U/c warning horn and light.	Operate test pushbutton.
81. Fire-extinguisher.	In position.
82. Forward upper escape hatch.	Operation. Secure.
83. Main parachute escape hatch.	Operation. Secure.
84. Front turret.	Condition. Doors closed.
85. Fire-extinguisher (in nose).	In position.
86. U/c selector lever.	Down. Safety bolt engaged.
87. Internal flying control locks.	Removed and stowed.
88. Pilot's seat.	Adjust for height.
89. Rudder pedals.	Adjust for length.
90. Flying controls.	Full and correct movement.

Cockpit checks.

N.B.—Work from left to right and then down the centre.

ITEM	CHECK
91. Port window and D.V. panel.	Operation.

ITEM	CHECK
92. Bomb door selector lever.	Up.
93. Air intake control.	Cold air.
94. Fuel jettison control.	Normal.
95. Navigation lights switch.	As required.
96. Auto pilot main switch (Mk. 4 auto pilot only).	Off.
97. Auto pilot control cock.	Spin.
98. Auto pilot clutches.	In. Engage controls.
99. Windscreen de-icing pump.	Operation.
100. Magnetic compass.	Serviceability.
101. External lights master switch.	As required.
102. U/c position indicator.	Switch on. (Mks. 1, 3 and 7). Operation.
103. Direction indicator.	Cage.
104. Identification lights.	As required.
105. D.R. compass.	Off.
106. Boost control cut-out lever.	Off.
107. No. 1 and No. 2 engine master cocks.	Off.

ITEM	CHECK	ITEM	ITEM
108. Landing lamps.	Operation. Retracted.	122. Elevator trimmer.	Full and correct movement.
109. D.R. compass repeater.	Synchronise.	123. Aileron trimmer.	Full and correct movement.
110. Auto pilot master switch (Mk. 8 auto pilot only).	Off.	124. Rudder trimmer.	Full and correct movement.
		125. Pilot's harness.	Adjust. Test lock.
111. Ignition switches.	Off.	126. Intercom.	Adjust headset. Test with crew.
112. Boost gauges.	Static readings.	127. Call lights.	Test with crew.
		128. Oxygen.	Delivery.
113. Flap indicator.	Switch on. Reading against position of flaps.	129. Entrance ladder.	Stowed.
		130. Entrance door.	Secured.
114. Supercharger switch.	Low gear.	131. Ground/ flight switch.	Ground (if necessary).
115. No. 3 and No. 4 engine master cocks.	Off.	**Start and warm up the engines** (see para. 38).	
116. Fuel cut-off switches (Mks. 3 and 10 aircraft).	Engine on.	132. Ground/ flight switch.	Flight.
		133. Engines.	Set to 1.200 r.p.m.
117. Pneumatic pressure gauge.	Available pressure. Pressure at each wheel brake.	134. Radiator shutters.	Open.
		135. D.R. compass.	On and setting.
118. Radiator shutters.	Automatic.	136. Fuel flow.	Check on all tanks with booster pumps on and off.
119. Starboard window and D.V. panel.	Operation.		
		137. Booster pumps.	Off.
120. Throttle control friction adjuster.	Function.	138. Vacuum pumps.	Changeover cock. Suction gauge readings.
121. R.p.m. control friction adjuster.	Function.	139. Flaps.	Operation. Return selector to neutral.

ITEM	CHECK
140. Pneumatic supply.	Pressure increasing to maximum.
141. Radio.	Test V.H.F. and other radio aids. Altimeter setting with control.
142. Altimeter.	Set.
143. Direction indicator.	Set with magnetic compass. Compare with D.R. compass. Uncage.
144. Engine temperatures and pressures.	Within limits.
145. Fuel flowmeters (if fitted).	Operation.
146. Generators.	Check during run-up.

Run up and test the engines (see para. 39).

Pilot's checks before and during taxying.

ITEM	CHECK
147. All hatches.	Secure.
148. Bomb doors.	Closed.
149. D.R. compass.	Normal.
150. Chocks.	Clear.
151. Taxying.	As soon as possible test brakes. Direction indicator for accuracy. Artificial horizon for accuracy. Brake pressure. Pressure head heater on if required.

Checks before take-off.

ITEM	CHECK
152. Trimming tabs—	
Elevator.	2 divs. nose heavy.
Rudder.	Neutral.
Aileron.	Neutral.

ITEM	CHECK
153. Throttle and r.p.m. controls friction dampers.	Tighten.
154. Superchargers.	Low gear.
155. Air intake control.	As required.
156. R.p.m. control levers.	Maximum r.p.m.
157. Fuel.	Contents. Master engine cocks on. Required tanks selected. Booster pumps on in Nos. 1 and 2 tanks. Fuel cut-off RUN (Mks. 3 and 10). Crossfeed cock off.
158. Flaps.	20° down.
159. Radiator shutters.	Automatic.
160. Direction indicator.	Set to magnetic compass and uncaged.
161. Auto pilot.	Clutches in. Control cock spin.
162. Engines.	Clear.
163. Harness.	Adjusted and locked.
164. Crew.	Warn.

Checks in flight as necessary.
Checks before landing.
When entering the circuit :—

ITEM	CHECK
165. Crew.	Warn.
166. Auto pilot.	Control cock spin.
167. Superchargers.	Low gear.

ITEM	CHECK	ITEM	CHECK
168. Air intake control.	As required.	182. Ignition switches.	Off.
169. Pneumatic supply.	Pressure sufficient. Delivery to each wheel brake.	183. Undercarriage indicator.	Off. (Mks. 1, 3 and 7).
170. Fuel.	Contents. Select fullest tanks. Booster pumps. on in Nos. 1 and 2 tanks. If either No. 1 or No. 2 tanks are empty keep their booster pumps off.	184. All fuel cocks.	Off.
		185. Flaps.	Select down. Indicator off. (Mks. 1, 3 and 7).
	Then reduce speed to 150 knots and check :—	186. Electrical services.	All off.
171. Flaps.	20° down.	187. Direction indicator.	Caged.
172. Undercarriage.	Down and locked (green lights on).	188. D.R. compass.	Off.
173. R.p.m. control levers.	Set as required. 2,850 r.p.m. on final approach.	189. Fuel cut-off switches (Mks. 3 and 10 aircraft.)	Run.
174. Flaps.	As required on final approach.	190. Chocks.	In position.
175. Harness.	Locked.	191. Brakes.	Off.
Checks after landing. When clear of landing area :—		192. Flying controls.	Locked
176. Radiator shutters.	Open.	193. Radiator shutters.	Automatic.
177. Flaps.	Up. Selector neutral.	194. Ground/ flight switch	Ground.
178. R.p.m. control levers.	Set to max. r.p.m. position.	195. Intercom.	Off.
179. Booster pumps.	Off.	196. Hydraulic accumulator.	Reading 220 lb./ sq. in.
180. Pressure head heater.	Off if necessary.	197. Nitrogen main cock (if fitted).	Off.
181. Brake pressure.	Sufficient for taxying.	198. Oxygen main cock.	Off.
On reaching dispersal. Idle the engines at 800-1,000 r.p.m. for a short period, test each magneto for a dead cut, then turn off the engine master cocks or operate the fuel cut-off switches or pushbuttons as applicable and when the engines have stopped :—		199. Static vents.	Plugs in.
		200. Pressure head.	Cover on.
		201. Form 700.	Sign if necessary.
		202. Authorisation book.	Sign.

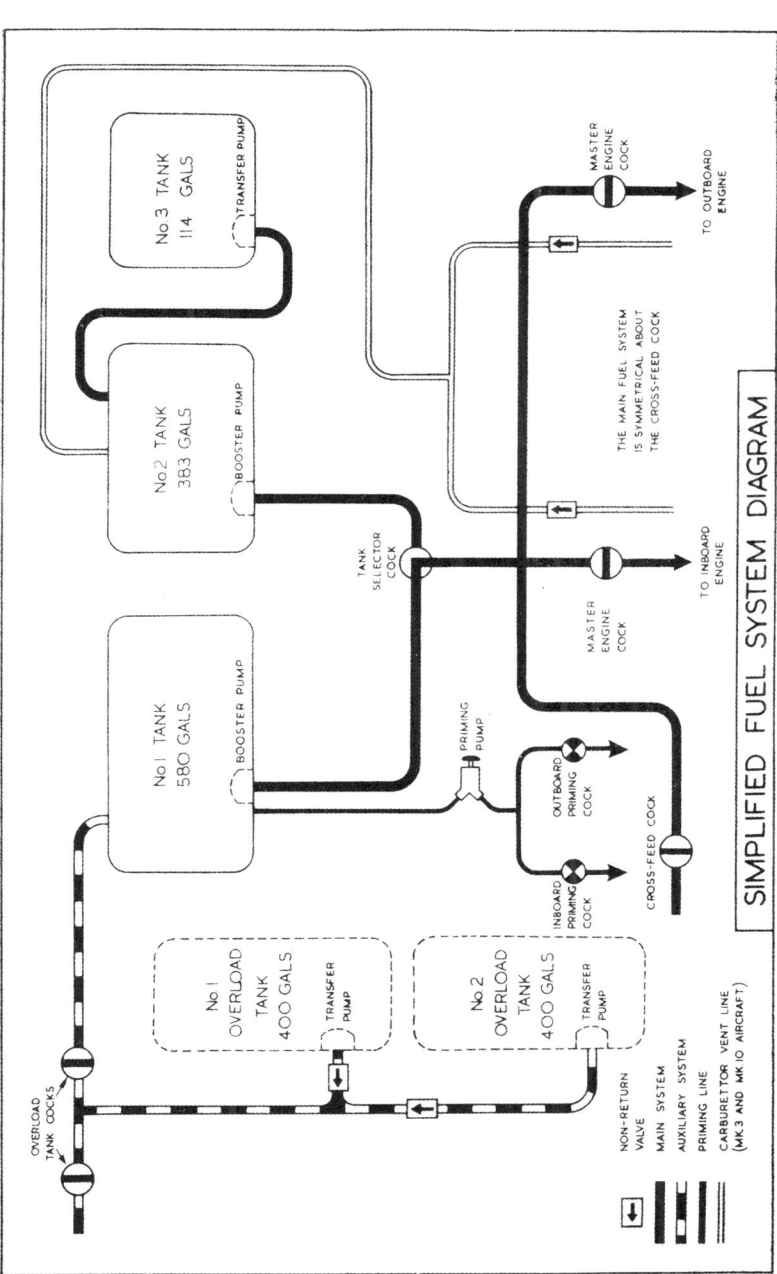

SIMPLIFIED FUEL SYSTEM DIAGRAM

PART I
DESCRIPTIVE

NOTE.—Throughout this publication the following conventions apply :—

(a) The numbers quoted in brackets after items in the text refer to the illustrations in Part V.

(b) Words in capital letters indicate the actual markings on the control concerned.

(c) Unless otherwise stated, all speeds quoted are indicated airspeeds.

INTRODUCTION

The main differences between the Lancaster Mks. 1, 3, 7 and 10 are in the power plants.

The Lancaster 1 is fitted with Merlin 20, 22 or 24 engines which have S.U. float-type carburettors.

The Lancaster 7 is fitted with Merlin 24 engines, and an electrically-operated mid upper turret.

The Lancaster 3 and 10 are fitted with Merlin 28, 38, or 224 engines which have Bendix-Stromberg pressure-injection carburettors. Lancaster 10 aircraft are Canadian built and differ from the British-built Lancasters in some of the instruments and in the electrical system.

Hydromatic 3-bladed propellers are fitted to all marks.

FUEL AND OIL SYSTEMS

1. **Fuel tanks**

(i) Three self-sealing fuel tanks are fitted in each wing, numbered 1, 2 and 3 outboard from the fuselage. The capacities are :—

Port and Starboard No. 1 ... 580 gallons each
Port and Starboard No. 2 ... 383 gallons each
Port and Starboard No. 3 ... 114 gallons each

1,077 gallons each side
2,154 gallons in all

The fuel from Nos. 1 and 2 tanks in each wing feeds through a tank selector cock directly to the engines on that side. The fuel in each No. 3 tank is transferred, when desired, into the corresponding No. 2 tank by a transfer pump. The fuel systems in each wing are independent, but are connected by a cross-feed pipe and cock.

(ii) Provision is made on some aircraft for carrying one or two 400-gallon tanks fitted in the bomb bay ; these tanks are connected so that their contents may be transferred into either or both No. 1 tanks and thence to the engines.

(iii) When the maximum bomb load is to be carried, the No. 2 tanks should be filled first, and the remainder of the fuel put in No. 1 tanks. This is on account of strength considerations of the aircraft structure.

2. **Fuel cocks**

(i) Four engine master cocks, (24) and (30), are fitted on the lower centre of the front panel, two on either side of the throttle quadrant. On Mks. 1 and 7 aircraft the engine master cocks also operate the slow-running cut-outs.

(ii) Two tank selector cocks (77) are fitted on the engineer's panel and control the fuel from Nos. 1 and 2 tanks in each wing.

(iii) A cross-feed cock, marked BALANCE COCK, is on the floor just forward of the front spar. It is reached through a hole in the spar cover and interconnects the port and starboard systems.

(iv) When the 400-gallon tanks are fitted in the bomb bay, they each have an ON-OFF cock situated behind the front spar in the centre of the fuselage.

3. **Vapour vent system** (Mks. 3 and 10 aircraft only).

A vent pipe from the two carburettors on one side of the aircraft is connected to the No. 2 tank on the same side, and allows vapour and a small quantity of fuel (approx. $\frac{1}{2}$ gal. per hour, per carburettor, but some carburettors may have a second vent allowing up to 10 gallons per hour) to return to the tank. This type of carburettor is designed to work full of fuel, and it therefore requires the vent to carry away any petrol vapour and dissolved air. The vent also assists in re-establishing the flow of fuel to the carburettors when the pipe-lines and pump have been run dry due to a tank emptying.

4. Nitrogen system

On some aircraft provision is made to allow nitrogen to be fed into the fuel tanks, as fuel is used, to avoid the risk of explosion should they be holed in flight. The control cock is on the starboard side of the fuselage between the front and rear spars and when it is required to use the system the cock should be turned fully on before take-off, and turned off after landing. A pressure gauge is fitted above the control cock.

5. Electric fuel booster and transfer pumps

A booster pump is fitted in each No. 1 and No. 2 tank, and a transfer pump is fitted in each No. 3 tank which is used to replenish the corresponding No. 2 tank by switching on the No. 3 tank pump. The pumps are controlled by switches (78) mounted on the engineer's panel ; the switches either have three positions, the down position being marked ON, the centre position OFF and the up position TEST, or else they have two positions, OFF and ON and a test pushbutton is mounted immediately above each switch. The TEST position of the switch (or the test pushbutton) is used in conjunction with an ammeter (68) on the engineer's panel for checking the pump operation. Similar switches control the pumps in the overload tanks (when fitted) to transfer their contents to the No. 1 tanks. The No. 3 tank switches are protected by guard covers, to prevent inadvertent operation.

6. Fuel gauges

(i) On Mks. 1, 3 and 7 aircraft, the switch (76) on the engineer's panel must be set ON before the fuel contents gauges (74) will indicate. On Mk. 10 aircraft there is no fuel contents gauge switch ; the gauges will indicate whenever electrical power is available.

16

(ii) Mods. 1198 and 1384 introduce " gallons gone " fuel flowmeters for each engine. Two combined indicators are mounted on the engineer's panel, and the counters can be set to zero by rotating anti-clockwise the knurled nut on the left-hand side of each indicator.

7. **Fuel pressure indicators**

Fuel pressure warning lights (79) on Mks. 1, 3 and 7 aircraft show when the fuel pressure at the carburettor falls appreciably below normal. They are switched off by the fuel contents gauges switch and this switch must, therefore, always be on in flight. On Mk. 10 aircraft, fuel pressure gauges (73) are fitted on the engineer's panel and indicate whenever electrical power is available.

8. **Priming system**

There is one cylinder priming pump in each inboard engine nacelle, drawing fuel from the No. 1 tank on that side ; each pump serves one inboard and one outboard engine. On Mks. 1, 3 and 7 aircraft this is accomplished by having two priming cocks fitted in each nacelle. On Mk. 10 aircraft the priming pump handle is turned to the left to prime the left engine, to the right to prime the right engine, and to the mid-position for off. Another cock and a short pipe may be fitted beside the priming pump and can be used to connect an outside supply of high volatility fuel for cold weather starting.

9. **Fuel jettisoning**

Originally fuel was able to be jettisoned if required from No. 1 tanks, and the control is on the floor to the left of the pilot's seat. This jettisoning system is now inoperative and the control should be kept in the normal (fully clockwise) position.

10. **Oil system**

Each engine has its own oil tank. The tanks, which are self-sealing, have a capacity of $37\frac{1}{2}$ gallons of oil with $4\frac{1}{2}$ gallons air space. An oil dilution system is fitted and is operated by pushbuttons (81) on the engineer's panel.

MAIN SERVICES

11. Hydraulic system

(i) Each turret is operated by an individual engine-driven
pump :
No. 1 engine ...	...	Tail turret
No. 3 engine ...	...	Front turret
No. 4 engine ...	...	Mid-upper turret (except on Mk. 7 aircraft where the mid-upper turret is electrically operated).

(ii) Two pumps (one each on No. 2 and No. 3 engines)
operate the following services through a small accumulator :—

Air-intake shutters.
Bomb-doors.
Flaps.
Undercarriage.

(iii) A handpump for ground-test purposes is mounted in the
fuselage, but owing to its small capacity, it is impracticable for use in flight.

(iv) The accumulator has an air charging valve and a pressure
gauge which should read 220 lb./sq. in. when there is no
hydraulic pressure in the system ; misleading pressure
gauge readings occur if the accumulator air pressure is
incorrect. The gauge should read between 800-900 lb./
sq. in. under working pressure when the cut-out operates,
isolating the pumps. The accumulator then provides the
initial pressure to operate the various systems. When
the pressure falls, the pumps will automatically be cut
in to operate the system and build up accumulator pressure again.

12. Pneumatic system

(i) A compressor on No. 3 engine charges an air bottle and
operates :—

Wheel brakes.
Radiator shutters.
Supercharger rams.
Fuel cut-off rams (on Mks. 3 and 10 aircraft only).

When Lincoln type undercarriage is fitted (Mod. 1195)

the air bottle is charged to 450 lb./sq. in. ; otherwise to 300 lb./sq. in.

A pressure-maintaining valve in the supply line from the air bottle only allows pressure to be supplied to the radiator shutters, superchargers and fuel cut-off rams, if the pressure in the air bottle exceeds 160 lb./sq. in. This is to ensure sufficient pressure for the brakes which operate at 125 lb./sq. in. It is necessary therefore to check on the pneumatic pressure gauge that the pressure is sufficient before high gear is engaged, or radiator override switches or fuel cut-off controls are operated.

(ii) A compressor on No. 2 engine operates the automatic pilot and the Mk. 14 bombsight. For the operation of the bombsight, the automatic pilot control cock must be set to OUT except on those aircraft in which Mod. 1161 is incorporated.

13. Vacuum system

A vacuum pump is fitted on No. 2 and No. 3 engines, one for operating the instruments on the instrument panel, and the other for operating the gyros of the Mk. 14 bombsight. The change-over cock (17) is on the right of the front panel beside the suction gauge (23), and in the event of failure of the vacuum pump supplying the flying instruments, the change-over cock can be used to connect the serviceable pump with the flying instruments, and cut out the bombsight. It is not possible to operate both the flying instruments and the bombsight on one vacuum pump.

14. Electrical system

(i) Two 1,500 watt generators are fitted on all aircraft. except Mk. 7 which have two 3,000 watt generators. The generators are fitted one on No. 2 and one on No. 3 engine and, connected in parallel, charge the aircraft batteries (24 volt) and supply the usual lighting and other services including :—

> Auto pilot.
> Bomb gear and bombsight.
> Camera.
> Dinghy release.
> D.R. compass.

Electro-pneumatic rams for :—
    Radiator shutters.
    Supercharger gear change.
    Air-intake emergency heat control.
    Fuel cut-offs (on Mk. 3 and Mk. 10 aircraft).
Engine starters and booster coils.
Fire-extinguishers.
Flaps position indicator.
Fuel booster and transfer pumps.
Fuel contents and flowmeter gauges.
Mid-upper turret (on Mk. 7 aircraft).
Oil dilution.
Propeller feathering.
Pressure-head heater.
Radio and radar.
Undercarriage position indicator.

An alternator may be fitted to Nos. 1 and 4 engines, to supply the radar equipment.

(ii) A ground/flight switch on the starboard side of the fuselage immediately aft of the front spar is used to isolate the aircraft batteries when the aircraft is parked or when using a ground starter battery.

(iii) The electrical control panel is fitted on the starboard side of the fuselage forward of the front spar, and mounted on the panel are a voltmeter, two generator switches, two ammeters and two earth warning lights. On Mk. 10 aircraft overload switches are fitted on the heavier circuits and flick off if the current becomes too great. If a switch flicks off it should be re-set only after the overloaded circuit has been allowed to cool for 30 seconds. If the switch flicks off again it indicates that the circuit is defective.

AIRCRAFT CONTROLS

15. **Flying controls**

The flying controls are conventional, and each rudder pedal is adjustable by holding aside the spring-loaded latch on each inside pedal arm and raising and moving the foot-rest over the ratchet mechanism.

16. **Flying controls locking gear**

The controls locking gear is stowed on the starboard side of the fuselage between the main spars, and consists of :—

(a) a strut to be fastened to the top of the pilot's seat and to a bracket on the control column.

(b) a strut to be inserted at one end into the cockpit left-hand rail, and fitted by two screwed hooks to the handwheel to prevent it rotating.

(c) a T-tube with a transverse member, to be inserted into the hollow footrest of each rudder pedal, and the other end attached to the bracket on the control column.

17. **Trimming tabs**

The elevator (62), rudder (63) and aileron (61) tab controls, on the right of the pilot's seat, all operate in the natural sense and each has an adjacent indicator showing the setting of the tab.

18. **Wheel brakes**

The brakes are applied by operating the lever (56) on the control column, and differential action is obtained by a relay valve connected to the rudder bar. The brakes may be locked for parking by compressing the lever fully and engaging the locking catch which is beside the lever. The available pneumatic pressure, and the pressure at each brake is shown on a triple pressure gauge (20) mounted on the right of the front panel.

19. **Undercarriage control**

The undercarriage lever (64) is locked in the DOWN position by a safety bolt (65) which has to be held aside in order to raise the lever. The bolt engages automatically when the lever is set to DOWN. The undercarriage may be lowered in an emergency by compressed air (see para. 61). There is no automatic lock other than the safety bolt, to prevent the undercarriage being raised by mistake when the aircraft is on the ground.

20. Undercarriage indicator

On Mks. 1, 3 and 7 aircraft, the indicator (39) shows as follows :—

Undercarriage locked down ... Two green lights
      ,,   unlocked ... ... Two red lights
      ,,   locked up ... ... No lights

The indicator switch (4) is interlocked so that it must be on when the No. 1 and No. 2 engine ignition switches are on. An auxiliary set of green lights can be brought into operation by pressing the central knob if failure of the main set is suspected. The red lamps are duplicated so that failure of one lamp does not affect the indication of undercarriage unlocked. The lights can be dimmed by turning the central knob.

On Mk. 10 aircraft a pictorial type of indicator is fitted. When the indicator switch is on and electrical power is available, the pictorial indicator shows the position of the undercarriage wheels and wing flaps. The disappearance of small red flags shows when the wheels are locked up or down.

21. Undercarriage warning horn

The horn sounds if either inboard throttle is closed when the undercarriage is not locked down. The outboard throttles do not operate the horn. A testing pushbutton and lamp are behind the pilot's seat on the cockpit port rail.

22. Flaps control

If the flaps have been selected partly down, and it is desired to lower them fully, it may be found that they will not lower further. This is due to the pressure in the accumulator having fallen below the pressure required to operate the flaps against the external air pressure but not sufficiently to cause the hydraulic pumps to cut in. To overcome this, move the selector (60) to UP, and then immediately put it fully DOWN ; it may be necessary to repeat this process more than once. This causes the hydraulic pumps to cut in. After the flaps have been selected fully down for landing, the selector should be left DOWN until landing is complete, to avoid any possibility of the flaps creeping up.

On Mks. 1, 3 and 7 aircraft, the position indicator (26) is switched on by a separate switch (27).

In an emergency the flaps may be lowered by compressed air after lowering the undercarriage (see para. 62).

23. **Bomb doors**

The control (43) has two positions only. The bomb release system is rendered operative soon after the doors begin to open and before they are fully open. The position of the doors must therefore be checked visually before releasing bombs. If the bomb doors open only part way and then stop, it is probably due to icing around the hinges and joints, which raises the hydraulic pressure sufficiently to bring the cut-out into operation, stopping any further movement of the doors. If the bomb doors selector is moved to SHUT and then immediately to OPEN, the doors will usually open further ; it may be necessary to repeat this several times to get the doors fully open. As strenuous pumping for 15 minutes is required to open the doors with engines stopped, they should be opened before stopping engines if the aircraft is to be bombed up before the next flight, or if necessary for servicing purposes.

24. **Automatic pilot**

On early aircraft the Mk. 4, and on later aircraft the Mk. 8 automatic pilot, is fitted, driven by a compressor in No. 2 engine.

ENGINE CONTROLS

25. **Throttle controls**

(i) *Merlin* 20, 22, 28 *and* 38 *engines.* Climbing boost, $+9$ lb./sq. in., is obtained with the throttle levers (28) at the gate. On Merlin 20 installations, going through the gate gives $+12$ lb./sq. in. at ground level only ; on Merlin 22, 28 and 38 engines it gives $+14$ lb./sq. in. When the boost control cut-out (32) is pulled, $+14$ lb./sq. in. is obtainable' in low gear, and $+16$ lb./sq. in. in high gear on all the above Merlins.

(ii) *Merlin* 24 *and* 224 *engines.* The throttle quadrant is fitted with a gate at $+9$ lb./sq. in boost ; the fully-forward position gives $+14$ lb./sq. in. at ground level only. When the boost control cut-out is pulled, $+18$ lb./sq. in. is obtainable in either gear.

26. Mixture control

(i) *Merlin* 20, 22 *and* 24 *engines.* (Mks. 1 and 7 aircraft.) S.U. float-type carburettors are fitted. The mixture strength is automatically controlled by boost pressure, and the pilot has no separate mixture control. A weak mixture is obtained below $+7$ lb. /sq. in. ($+4$ lb./sq. in. on Merlin 20). The carburettor slow-running cut-outs are operated by closing the engine master cocks.

(ii) *Merlin* 28, 38 *and* 224 *engines.* (Mks. 3 and 10 aircraft.) Bendix-Stromberg pressure injection carburettors are fitted. There is no pilot's mixture control, the mixture strength being regulated by the boost so that an economical mixture is obtained below $+7$ lb./sq. in. The fuel cut-offs, which are used in starting and for stopping the engines, are operated by electro-pneumatic rams controlled by four two-position switches (11) or four push-buttons (if Mod. 1753 is fitted) mounted on the pilot's panel above the engine starter buttons. When two-position switches are fitted the top position is the ENGINE RUN position, and the bottom position is the IDLE-CUT-OFF position ; when pushbuttons are fitted, they have to be held in to keep them in " cut-off."

NOTE.—(a) If the pneumatic supply pressure is less than 160 lb./sq. in., it is possible to start the engines with the fuel cut-off switches in the IDLE-CUT-OFF position ; then, when the supply pressure builds up the cut-off rams will operate and all four engines will stop.

(b) When the aircraft is parked the fuel-cut-offs should be left in the ENGINE RUN position. If left in the IDLE-CUT-OFF position the rams will return to the " engine run" position when the electrical current is switched off, and back to " cut-off " when it is switched on again, with a consequent waste of pneumatic pressure.

24

PART I—DESCRIPTIVE

27. Propeller controls

(i) The r.p.m. control levers (29) are mounted in a quadrant on the engine controls pedestal and vary the governed r.p.m. from 3,000 down to 1,800.

(ii) The feathering pushbuttons (19) are mounted on the right of the front panel and when Mods. 1067 and 1314 are fitted, engine fire warning lights are mounted in the respective feathering pushbuttons (see paras. 54 and 69).

28. Superchargers control

The superchargers gear-change is electro-pneumatically operated by a switch on the front panel immediately below the engine speed indicators. The switch, which has two positions, MS and FS, controls all four engines simultaneously and a red warning light (25) beside it indicates if FS gear is engaged when the main wheels are down. In the event of electrical or pneumatic failure, the rams will stay at, or return to, the MS gear position.

29. Carburettor air-intake heat control

A single lever for the hydraulic operation of all four carburettor's warm air intakes is on the left of the pilot's seat.

When Mod. 1198 is fitted the control has 3 positions, COLD, WARM and HOT (emergency). If the control is at WARM, air is then drawn into each engine through the warm air-intake inside the bottom engine cowling. When moved to HOT, a flap in the bottom engine cowling is opened by an electro-pneumatic ram and allows hot air to be drawn from the radiator to the engine. The WARM position may be used to prevent the formation of ice when flying in icing conditions, but this will reduce the range (see para. 52 (iii)). If, instead, the flight is continued in COLD until carburettor icing becomes evident, the HOT position should then be used, but only until the ice has cleared.

If Mod. 1198 is not fitted, there are only two positions COLD and WARM, but the latter should be regarded as having a similar function to the HOT position when the 3-position control is fitted, as air is drawn from the radiator to the engine through a hole in the engine cowling when the control is at WARM.

30. **Radiator shutters**

The radiator shutters are controlled by two-position switches marked AUTOMATIC and OPEN, and are mounted on the cockpit right-hand wall. The shutters are automatically operated by a thermostat when they are set to the AUTOMATIC position ; when the switches are in the OPEN position the thermostatic control is overriden, and this position should be used for all ground running and taxying.

When the aircraft is parked, the shutters should be left in the AUTOMATIC position. If left in the OPEN position, they will close when the electrical current is switched off, and open when it is switched on again, with a consequent waste of pneumatic pressure.

31. **Ignition and starting controls**

The ignition switches (7), booster-coil master switch (10) and the starter pushbuttons (14) are all at the top centre of the front panel. The starter pushbuttons operate the booster coils when the booster-coil switch is ON.

OTHER EQUIPMENT

32. **Cockpit lighting**

(i) Lighting is provided by two floodlights, which are controlled by adjacent dimmer switches in the cockpit roof. The engineer's panel is provided with a separate light and dimmer switch.

(ii) On some aircraft a single emergency light is fitted for use only in the event of complete electrical failure. It is powered by a separate small battery and is controlled by a switch on the left-hand side of the front panel.

33. **Cockpit heating and ventilation**

The cockpit is heated by warm air from the two radiators mounted in the mainplane leading edge and connected to the inboard engine cooling systems. On each side of the fuselage, just forward of the front spar, is a control knob which operates a shutter in the air duct. When turned anti-clockwise it opens the inlet to the cabin and closes the by-pass to the outer air. To control the escape

FINAL CHECKS FOR TAKE-OFF

TRIM	... **ELEVATORS : 2 DIVS. NOSE HEAVY**
	RUDDER : NEUTRAL
	AILERON : NEUTRAL
PROPELLERS	... **MAX. R.P.M.**
FUEL	... **MASTER COCKS ON**
	CORRECT TANKS SELECTED
	NOS. I AND 2 TANKS BOOSTER PUMPS ON
	CROSSFEED COCK OFF
	FUEL CUT-OFFS "RUN"
FLAPS	... **20° DOWN**
SUPER-CHARGERS	... **LOW GEAR**
AUTO-PILOT	... **SPIN**

FINAL CHECKS FOR LANDING

FUEL ... CHECK CONTENTS

CORRECT TANKS
SELECTED

NOS. 1 AND 2 TANKS
BOOSTER PUMPS ON

SUPER-
CHARGERS ... LOW GEAR

BRAKES ... OFF
CHECK PRESSURES

UNDER-
CARRIAGE ... DOWN AND LOCKED

PROPELLERS ... 2,850 R.P.M. ON FINAL

FLAPS ... AS REQUIRED

of the air from the cabin an extractor louvre is provided on each side of the nose of the fuselage.

34. **Pilots' seats**

The left-hand pilot's seat is provided with hinged arm-rests and is adjustable for height by a lever on the left-hand side. The right-hand pilot's seat is collapsible and folds up against the side of the fuselage, being secured by a strap to the cockpit rail when not in use. A back-rest is positioned above the seat and can be folded up against the fuselage side when not required. A bar is provided as a foot-rest and, when not in use, slides under the left seat platform.

35. **Windscreen de-icing**

Two de-icing sprays for the windscreen are operated by a handpump (59) on the floor, forward and to the left of the pilot's seat. The pump is operated by pressing down the handle which is spring-loaded to return at a pre-set speed according to the setting of the needle valve of the pump. A setting of $1\frac{2}{3}$ turns on the needle valve is recommended when the pump, operated approximately once a minute, will give a delivery at the rate of two pints per hour. The tank, which also supplies the bomb-aimers' panel, is of approximately 4 gallons capacity and is fitted on the starboard side between the nose and the front centre section of the aircraft.

36. **Oxygen system**

The pilot's flexible oxygen pipe (51) is secured by spring clips to the cockpit left-hand rail, and the economiser is below the rear end of the pilot's floor. A Mk. 8 series regulator (18) which controls the supply throughout the aircraft is fitted on the right of the front panel.

PART II

HANDLING

37. **Management of the fuel system**

(i) *Use of tanks.* Structural considerations render it advisable that fuel should be kept outboard as much as possible.

(a) Start and warm up on No. 1 tanks. Run up and take-off on No. 2 tanks, and continue to fly on these tanks for the first hour of flight. This allows space for carburettor venting, where applicable (see para. 3).

(b) After the first hour of flight No. 1 tanks should be selected. When nearly empty, transfer the fuel from the fuselage overload tanks, if in use, by turning on both long-range fuel cocks (behind the front spar) and switching on the overload tank pump switches. The fuel contents gauges switch must also be on. Transfer of fuel from the long-range tanks takes approximately one hour. Turn off each long-range tank cock and the appropriate pump switch when the tank is empty.

(c) Continue to run on No. 1 tanks until empty ; then re-select No. 2 tanks and use until approximately 200 gallons remain in each. Then transfer the contents of No. 3 tanks by switching on the transfer pumps.

Switch off the pumps when No. 3 tanks are empty.

(ii) *Use of the booster pumps*

The main use of the booster pumps in No. 1 and No. 2 tanks is to maintain fuel pressure at altitudes of approximately 17,000 ft. and over in temperate climates, but they are also used for raising the fuel pressure before starting and to assist in re-starting an engine during flight. If one engine fails and the booster pump is not ON, air may be drawn back into the main fuel system before the engine master cock of the failed engine can be closed, thus causing the failure of the other engine on the same side. At take-off, therefore, the pumps in Nos. 1 and 2 tanks must be switched on ; this is also a precaution against fuel failure during take-off as an immediate supply is available by changing over the tank selector cock. The pump in each tank in use should also be switched on at any time when a drop in fuel pressure is indicated or when it is necessary to run all engines from one tank by opening the cross-feed cock.

(iii) *Testing the booster and transfer pumps*

Before starting the engines, each booster and transfer pump should be tested by the ammeter fitted on the engineer's panel ; to do this the switch for each pump (on Mks. 1, 3 and 10 aircraft) should in turn be set to the up (TEST) position, after ensuring that the engine master cocks are OFF. On Mk. 7 aircraft the pumps are tested by pressing the test pushbuttons above the switches. The ammeter reading should in all cases be perfectly steady and should be between 4 and 7 amps. on Mks. 1, 3 and 7 aircraft, and between 3 and 5 amps. on Mk. 10 aircraft.

(iv) *Use of cross-feed cock*

The cock should be closed at all times, unless it is necessary in an emergency to feed fuel from the tanks in one wing to the engine(s) in the other wing. If the cross-feed cock is open for this purpose, select the tank from which it is desired to cross-feed fuel, switch on the pump in this tank, and turn off the selector cock of the opposite wing tanks.

38. **Starting the engines and warming up**

NOTE.—On Mks. 3 and 10 aircraft the fuel booster pumps must never be switched on with the engine master cock open and the engine stationary, unless the fuel cut-off switch, or pushbutton, is in the IDLE-CUT-OFF position, and the pneumatic supply pressure not less than 160 lb./sq. in.

(i) *Preliminaries.* After carrying out the external, internal and cockpit checks, laid down in the Pilot's Check List, ensure :—

Engine master cocks	...	OFF
Fuel cut-off controls	...	ENGINE RUN
Throttles	...	$\frac{1}{2}$ inch open
R.p.m. controls ...	...	Control levers max. r.p.m.
Superchargers control	...	MS gear (warning light out)
Air intake heat control	...	COLD
Radiator shutters...	...	Override switches at
		AUTOMATIC
Booster pumps ...	...	OFF
Booster-coil master switch		ON

(ii) Turn on the engine master cock of the engine to be started, and if the aircraft is to be started from external batteries, set the ground/flight switch to GROUND.

(iii) Prime the carburettor of the engine to be started by putting the fuel cut-off to IDLE-CUT-OFF (on Mks. 3 and 10 aircraft) and switching the booster pump in the No. 1 tank on for a period of 10 seconds. Switch off the booster pump and then return the fuel cut-off to the ENGINE RUN position.

(iv) High volatility fuel should be used if an outside priming connection is fitted, for priming at air temperatures below freezing. The ground crew should work the priming pump until the fuel reaches the priming nozzles ; this may be judged by an increase in resistance.

(v) Switch on the ignition and press the starter button. Turning periods must not exceed 20 seconds with a 30 second wait between each. The ground crew should work the priming pump as firmly as possible while the engine is being turned.

(vi) It will probably be necessary to continue priming after the engine has fired, and until it picks up on the carburettor. When the engine is running smoothly, proceed to prime the carburettors and start the other engines in turn.

(vii) When all the engines are running satisfactorily, switch off the booster-coil master switch. The ground crew should screw down the priming pumps and turn off the priming cocks (if fitted).

(viii) Ensure that the ground/flight switch is turned to FLIGHT and have the external battery removed if used.

(ix) Open up each engine slowly to 1,200 r.p.m. and warm up at this speed.

(x) While warming up carry out the checks detailed in the Pilot's Check List, items 132 to 146.

39. **Exercising and Testing**

(i) After warming up until the oil temperature is $+15°C$. and the coolant temperature is $+40°C$., switch the radiator shutters over-ride switches to OPEN, turn the tank selector cocks to No. 2 tanks, and test each magneto as a precautionary check before increasing power further. Then for each engine in turn :—

    (a) Open up to the static boost reading (0 lb./sq. in. under " standard atmosphere " conditions) and exercise and check the operation of the constant speed propeller by moving the lever over its full range at least twice. With the lever fully forward check that the r.p.m. are within 50 of those normally obtained.

    (b) At the same boost check the operation of the supercharger. R.p.m. should fall, boost rise and the supercharger warning light should come on when high gear is engaged. Change back to low gear and ensure that the original conditions are restored.

    (c) At the same boost check with the engineer that the generators on Nos. 2 and 3 engines are charging.

    (d) At the same boost test each magneto in turn. If the single ignition drop exceeds 150 r.p.m., but there is no undue vibration, the ignition should be checked at higher power, see below ; if there is marked vibration the engine should be shut down and the cause investigated.

    NOTE.—The following full power checks should be carried out after repair, inspection other than daily, when the single ignition drop at the static boost reading exceeds 150 r.p.m. or at the discretion of the pilot. Except in these circumstances if the checks above are satisfactory no useful purpose will be served by a full power check.

    (e) Open the throttle fully and check take-off boost and r.p.m. This check should be as brief as possible.

    (f) Throttle back until the r.p.m. fall just below the take-off figure and test each magneto in turn. If the single ignition drop exceeds 150 r.p.m. the aircraft should not be flown.

(g) After completing the checks, either at the static boost reading, or at full power, steadily move the throttle to the fully-closed position, and check the minimum idling r.p.m. then open up to between 1,000 and 1,200 r.p.m.

(ii) Before and during taxying carry out the checks in the Pilot's Check List, items 147 to 151.

40. Taking off

(i) After carrying out the checks, items 152 to 164 laid down in the Pilot's Check List, clear the engines by opening up to the static boost reading if the run-up has not been carried out immediately prior to taxying on to the runway.

(ii) Align the aircraft carefully on the runway, making certain that the tailwheel is straight.

(iii) Release the brakes and open the throttles slowly to the take-off position.

(iv) Keep straight by coarse use of the rudder and by differential throttle opening.

(v) As speed is gained, ease the control column forward to raise the tail. Do not attempt to raise the tail by exerting a heavy push force on the control column during the early stages of the take-off run.

(vi) At 65,000 lb. ease the aircraft off the ground at 90 knots and at 72,000 lb. at 105 knots.

(vii) When comfortably airborne, brake the wheels and retract the undercarriage.

(viii) With flaps 20° down, safety speed at 65,000 lb. is 145 knots when using +18 lb./sq. in. boost and 3,000 r.p.m. ; at 72,000 lb. it is 150 knots. In view of these high speeds power should, where practicable, be reduced early after take-off.

(ix) At a safe height raise the flaps in stages. Then return the selector to neutral.

(x) The booster pumps in Nos. 1 and 2 tanks may be switched off after the initial climb, but if a warning light comes on (or on Mk. 10 aircraft the fuel pressure gauge shows less than 10 lb./sq. in.), switch on No. 2 pumps immediately.

41. **Climbing**

The speed for maximum rate of climb is 140 knots, but to improve control when climbing at full load speeds may be increased to 155 knots.

42. **General flying**

(i) *Stability.*—At normal loadings and speeds stability is satisfactory. At loads above 67,000 lb. there is a tendency for the aircraft to wallow. It is not advisable to attempt to correct this as use of the controls may aggravate it.

(ii) *Controls.*—The elevators are relatively light and effective, but tend to become heavy in turns. The ailerons are light and effective but become heavy at speeds over 225 knots, and also at heavy loads. The rudders also become heavy at high speeds.

(iii) *Changes of trim :—*

Undercarriage UP	Slightly nose up
Undercarriage DOWN ...	Slightly nose down
Flaps up to 20° from fully DOWN	Slightly nose down
Flaps up from 20°	Strongly nose down
Flaps down to 20°	Strongly nose up
Flaps fully DOWN from 20°	Slightly nose up
Bomb doors open	Slightly nose up

(iv) *Flying at reduced airspeeds.*—Flaps may be lowered about 20°, r.p.m. set to 2,650, and the speed reduced to about 115 knots.

43. **Maximum performance**

(i) *Climbing :—*
140 knots to 12,000 ft.
135 knots from 12,000 to 18,000 ft.
130 knots from 18,000 to 22,000 ft.
125 knots above 22,000 ft.
Change to high gear when boost has fallen to $+6$ lb./sq. in.

(ii) *Operational necessity*
Use high gear if the boost obtainable in low gear is more than 3 lb./sq. in. (4 lb./sq. in. with **Merlin** 24 or 224) below maximum boost.

44. Maximum range

(i) Climbing.—140 knots at +7 lb./sq. in. boost with Merlin 22, 24, 28, 38 or 224 (+4 lb./sq. in with Merlin 20) and 2,650 r.p.m. Change to high gear when maximum boost obtainable in low gear has fallen by 3 lb./sq. in.

(ii) Cruising (including descent) :—

(a) Fly in low gear at maximum obtainable boost not exceeding +4 lb./sq. in. with Merlin 20, +7 lb./sq. in. with Merlin 22, 24, 28, 38 or 224 obtaining the recommended airspeed by reducing r.p.m., which may be as low as 1,800 if this will give the recommended speed. Higher speeds than those recommended may be used if obtainable in low gear at the lowest possible r.p.m.

(b) The recommended speeds are :—

Fully loaded
Up to 15,000 ft.	150 knots
At 20,000 ft. in high gear	140 knots
Lightly loaded	140 knots

(c) Engage high gear when the recommended speed cannot be maintained at 2,500 r.p.m. in low gear.

(iii) The use of warm air intakes will reduce air nautical miles per gallon considerably. On this installation there is no need to use warm air unless intake icing is indicated by a drop in boost.

45. Fuel capacity and consumptions

(i) Capacity :—

Two No. 1 tanks	1,160 gallons
Two No. 2 tanks	766 gallons
Two No. 3 tanks	228 gallons
Normal total ...	2,154 gallons
Two 400 gallon overload tanks ...	800 gallons
Overload total ...	2,954 gallons

(ii) Weak mixture consumptions, Merlin 20, 22 or 24 :—
The following figures are the approximate total gallons per hour and apply in low gear between 8,000 and 17,000 feet, and in high gear between 14,000 and 25,000 feet.

Boost lb./sq. in	R.p.m.		
	2,650	2,300	2,000
+7*	260*	225*	212*
+4	228	204	188
+2	212	188	172
0	192	172	150
−2	172	156	140
−4	152	136	124

* These figures do not apply to Merlin 20.

(iii) Weak mixture consumptions Merlin 28, 38 and 224 engines :—
The following figures are the approximate total gallons per hour for the aircraft and apply in low gear at 5,000 ft. and high gear at 15,000 ft. One gallon per hour should be added for every 1,000 ft. above these heights.

Boost lb./sq. in.	R.p.m.				
	2,650	2,400	2,200	2,000	1,800
+7	240	235	217	200	—
+4	216	204	196	180	—
+2	196	184	176	164	—
0	172	164	156	144	128
−2	148	140	128	124	112
−4	124	120	108	104	96

(iv) Rich mixture consumptions, Merlin 20, 22, 24.

Boost lb./sq. in.	R.p.m.	Total gallons per hour
+14	3,000	500
+12	3,000	460
+9	2,850	380
+7*	2,650*	320*

* Merlin 20 only.

(v) Rich mixture consumptions, Merlin 28, 38 and 224 :—

Boost lb./sq. in.	R.p.m.	Total gallons per hour
+9	2,850	420

46. Position error corrections

(i) The position error correction is −1 knot at 'all speeds from 120 knots upward.

(ii) When an H_2S blister is fitted, the position error corrections are as follows :—

From	110	130	160	190	knots
To	130	160	190	250	knots
Subtract	1	2	3	4	knots

(iii) On Lancaster ASR Mk. 3 the corrections are as follows :

    (a) When the lifeboat is being carried, the correction is +3 knots at speeds below and +2 knots at speeds above 140 knots.

    (b) When the lifeboat is not being carried the correction varies from +1 knot at 95 knots to −1 knot at 215 knots.

47. Stalling

(i) Warning of the stall is given by slight tail buffeting, which generally commences some 5 knots before the stall itself.

At the stall the nose drops gently. Recovery is straight-forward and easy.

(ii) The approximate stalling speeds are :—

	At 60,000 lb.	At 65,000 lb.
Power off		
Undercarriage and flaps up	95	105
Undercarriage and flaps down	80	85
Power on (approach conditions, −2 lb./sq. in. boost, 2,850 r.p.m. set)		
Undercarriage and flaps down	70–75	—

(iii) *High speed stall.*—Adequate warning of the approach of a stall in a turn is given by strong rudder and elevator buffeting. At the stall the inner wing and nose drop gently together. Recovery is immediate on pushing the control column forward.

48. Diving

There is a strong nose-down change of trim as speed is gained in the dive. On aircraft in which Mod. 1101 or 1131 is incorporated it is possible to recover from dives to the limiting speed without the assistance of the elevator trimming tab, even if the aircraft has been trimmed into the dive. If the elevator trimming tab is used it should be applied with care, since it is powerful and sensitive. On unmodified aircraft the elevator trimming tab should never be used to assist entry, but should be used to reduce the very heavy pull force otherwise necessary for recovery.

49. Approach and landing

After carrying out the checks, items 165 to 175, laid down in the Pilot's Check List, the turn into wind should be made at about 115 knots, airspeed being reduced progressively, so that the airfield boundary is crossed at the following speeds :—

	At light load 45,000 lb.	At max. landing weight 60,000 lb.
Engine assisted Flaps down	95	105

50. Mislanding and going round again

(i) At maximum landing weight the aircraft will climb away satisfactory with the undercarriage and flaps down at maximum climbing power.

(ii) After increasing power, maintain an airspeed of 125 knots, select undercarriage up, and while the undercarriage is rising, select the flaps up to 30° and thereafter in stages, retrimming as necessary.

51. After landing

(i) Carry out the checks in the Pilot's Check List, items 176 to 181.

(ii) Before stopping the engines, open the bomb doors if required for bombing up or servicing purposes.

(iii) *Stopping engines*

   (a) *Merlin* 20, 22 *and* 24
   With the engines running at 800 r.p.m. turn OFF the engine master cocks and switch OFF the ignition after the engines have stopped.

   (b) *Merlin* 28, 38 *and* 224
   Check that the pneumatic pressure gauge reads at least 160 lb./sq. in. If not, open up No. 3 engine to increase the pneumatic pressure, and then, with the engine running at about 800 r.p.m., move the fuel cut-off to the " cut-off " position.
   Do not stop the engines by turning off the engine master cocks, as this will empty the carburettors of fuel and entail trouble in any subsequent starting attempt. When all the engines have stopped, switch off the ignition and turn off the engine master cocks.

(iv) Carry out the checks in the Pilot's Check List, items 182 to 202.

(v) *Oil dilution*

   The recommended dilution period for this aircraft is :—
   Air temperatures above $-10°C.$ one minute
   ,,   ,   below $-10°C.$ two minutes

PART III

LIMITATIONS

52. Engine data

(i) *Merlin* 20

Engine limitations with 100/130 grade fuel :—

	Gear	R.p.m.	Boost lb./sq. in.	Temp. °C. Coolant	Oil
MAX. TAKE-OFF 5 MINS. LIMIT	Low	3,000	+12*		
MAX. CLIMBING 1 HOUR LIMIT	Low } High }	2,850	+9	125	90
MAX. RICH CONTINUOUS	Low } High }	2,650	+7	105	90
MAX. WEAK CONTINUOUS	Low } High }	2,650	+4	105	90
OPERATIONAL NECESSITY 5 MINS. LIMIT	Low High	3,000 3,000	+14 +16	135 135	105 105

* +14 lb./sq. in. is obtained by operating the boost control cut-out.

OIL PRESSURE :
  MINIMUM IN FLIGHT   ...   30 lb./sq. in.
MINIMUM TEMPS. FOR TAKE-OFF :
  OIL   ...   .   +15°C.
  COOLANT   ...   .   +40°C.

(ii) *Merlin* 22, 28 *or* 38

Engine limitations with 100/130 grade fuel :—

	Gear	R.p.m.	Boost lb./sq. in.	Temp. °C. Coolant	Oil
MAX. TAKE-OFF 5 MINS. LIMIT	Low	3,000	+14		
MAX. CLIMBING 1 HOUR LIMIT	Low } High }	2,850	+9	125	90
MAX. CONTINUOUS	Low } High }	2,650	+7	105	90
OPERATIONAL NECESSITY 5 MINS. LIMIT	Low High	3,000 3,000	+14 +16	135 135	105 105

OIL PRESSURE :

MINIMUM IN FLIGHT   ...   30 lb./sq. in.

MINIMUM TEMPS. FOR TAKE-OFF :

OIL ...   .   +15°C.

COOLANT ...   .   +40°C.

(iii) *Merlin* 24 *or* 224

Engine limitations with 100/130 grade fuel :—

	Gear	R.p.m.	Boost lb./sq. in.	Temp. °C. Coolant	Oil
MAX. TAKE-OFF 5 MINS. LIMIT	Low	3,000	+18*		
MAX. CLIMBING 1 HOUR LIMIT	Low } High	2,850	+9	125	90
MAX. CONTINUOUS	Low } High	2,650	+7	105	90
OPERATIONAL NECESSITY 5 MINS. LIMIT	Low } High	3,000	+18*	135	105

* +18 lb./sq. in. boost must not be used below 2,850 r.p.m.

OIL PRESSURE :

MINIMUM IN FLIGHT   ...   30 lb./sq. in.

MINIMUM TEMPS. FOR TAKE-OFF :

OIL ...   .   +15°C.

COOLANT ...   .   +40°C.

53. **Flying limitations**

(i) The aircraft is designed for manœuvres appropriate to a heavy bomber and care must be taken to avoid imposing excessive loads with the elevators in recovery from dives and in turns at high speed. Violent use of the rudder at high speeds should be avoided.

(ii) Maximum speeds in knots :—

Diving	315
Bomb doors open	315
Undercarriage down	175
Flaps down	175

(iii) Maximum weights :—

Take-off, straight flying and gentle manœuvres 63,000 lb.

,, ,, ,, ,, ,, ,, †64,000 lb.

,, ,, ,, ,, ,, ,, ‡65,000 lb.

,, ,, ,, ,, ,, *72,000 lb.

Landing and all forms of flying ... ... 55,000 lb.

Landing ... ... *60,000 lb.

†This weight is permitted for Lancaster ASR. Mk. 3.

‡This weight is permitted provided the following mods. are incorporated : Mod. 503 or 518, Mod. 588 or 598, Mod. 811 or SI/RDA. 600 and Mod. 1004.

*These weights are permitted if Merlin 24 or 224 power plants are fitted, paddle-bladed propellers are fitted. Lincoln type undercarriage and tyres (Mod. 1195) are fitted and special adjustments are made to the tyre and oleo leg pressures. A careful check on aircraft structure must be kept and runways only must be used.

PART IV
EMERGENCIES

54. Feathering

(i) Close the throttle.

(ii) Press the feathering pushbutton and hold it in only long enough to ensure that it stays in by itself. Then release it and check that it springs out when feathering is complete. If it does not do so, it must be pulled out by hand.

(iii) Switch off the ignition when the propeller has stopped (or nearly stopped) rotating, and turn off the engine master cock at once.

(iv) If the engine has been feathered because of fire, operate the engine fire-extinguisher as soon as the propeller has stopped turning (see also para. 69).

(v) Engine auxiliaries which will be affected by feathering :—

No. 1 engine.—Alternator for radar, rear turret hydraulic pump.

No. 2 engine.—Generator, main services hydraulic pump, compressor for automatic pilot and computor unit of Mk. 14 bombsight, No. 1 vacuum pump.

No. 3 engine.—Generator, main services hydraulic pump, front turret hydraulic pump, compressor for pneumatic system, No 2 vacuum pump.

No. 4 engine.—Alternator for radar, mid-upper turret hydraulic pump (when fitted).

55. Unfeathering

(i) Switch on the ignition, set the throttle as for starting and the r.p.m. control lever to the minimum r.p.m. position.

(ii) Check that the fuel booster pump of the tank in use is OFF, then press the feathering pushbutton. As the engine starts to turn, set the engine master cock to ON. Continue to hold the button in until r.p.m. reach 800-1,000 Check that the pushbutton springs out when released ; if it does not do so it must be pulled out by hand.

(iii) If the propeller does not return to normal constant-speed operation it must be refeathered and then unfeathered again, releasing the button at slightly higher r.p.m.

56. Engine failure during take-off

(i) With flaps 20° down safety speed at 65,000 lb. is 145 knots, using + 18 lb./sq. in. boost and 3,000 r.p.m.

(ii) If engine failure occurs below critical speed it will always be necessary to throttle back the opposite outer engine at least partially to maintain control.

(iii) With the propeller of the failed engine feathered and rudder trim applied to relieve footload it should be possible for the aircraft to climb away at 125 knots at weights up to 65,000 lb.

(iv) As soon as the undercarriage is up, raise the flaps in stages, retrimming as necessary.

57. Handling on three engines

The aircraft will maintain height at loads up to 65,000 lb. on any three engines at 10,000 feet and can be trimmed to fly without footload. Maintain at least 130 knots.

58. Landing on three engines

Lowering of flaps to 20° and of undercarriage may be carried out as normally on the circuit but further lowering of the flaps should be left until the final straight approach.

59. Going round again on three engines

The decision to go round again should be made before full flap is lowered. With the flaps lowered 20° and the undercarriage down, power should be increased to +9 lb./sq. in. boost, 2,850 r.p.m. The aircraft can be controlled comfortably at 130 knots. Select undercarriage up and while it is rising, select flaps up in stages, retrimming as necessary.

60. Flying on asymmetric power on two engines

(i) *In flight.*—It should be possible to maintain height below 10,000 feet at 125 knots on any two engines after release of bombs and with half fuel used, but with two engines dead on one side, the footload will be very heavy.

(ii) *Landing.*—A circuit in either direction can safely be made irrespective of which engines have failed. While manœuvring with the undercarriage and flaps up maintain a speed of at least 130 knots. Aim to have the undercarriage locked down at the end of the downwind leg. The flaps should not be lowered until the final approach is commenced and it is certain that the airfield is within easy reach. The two live engines should be used within the limits of rudder control to regulate the rate of descent. Speed and power should gradually be reduced, aiming to cross the airfield boundary at the normal engine-assisted approach speed.

61. Undercarriage emergency operation

If the hydraulic system fails, the undercarriage can be lowered by compressed air from a special bottle or bottles, irrespective of the position of the undercarriage lever.

The flap selector should be neutral before using the emergency air system.

The knob (80) for working the air system is just forward of the engineer's panel. The undercarriage cannot be raised again by this method. Although it will lower by this method irrespective of the position of the normal selector, the latter must be selected DOWN for landing before operating the emergency air system, and left in the down position after landing ; otherwise any leakage of air pressure may cause the locks to be released and the undercarriage to collapse.

62. Flaps emergency operation

After lowering the undercarriage by turning on the emergency air cock, the flaps may be lowered by operating the flaps control, which admits the air pressure to the flaps system. The flaps can be raised again, but there may not be sufficient air pressure to lower the flaps a second time ; furthermore it may cause the header tank to burst. If it is absolutely necessary to raise the flaps by the emergency method extreme care must be taken to raise them slowly by stages. If the flaps are lowered by the emergency method before landing they must be left down after landing, owing to the likelihood of bursting the header tank, if they are raised.

63. Flapless landings

The initial approach should be made at 120 knots ; little power will be required to maintain this speed. The approach is flat with a nose-up attitude, but control remains satisfactory. Aim to cross the airfield boundary at :—

45,000 lb	60,000 lb.
100 knots	115 knots

The touchdown is straightforward and the aircraft can easily be brought to rest within 2,000 yards.

64. **Bomb jettisoning**

(i) Open the bomb doors and check visually that both are fully open. See para. 23.

(ii) Then jettison the containers first by the switch (15) on the right of the front panel.

(iii) Jettison the bombs by the handle (16) beside the container jettison switch.

(iv) Close the bomb doors.

65. **Parachute exits**

The hatch in the floor of the nose should be used by all members of the crew if time is available ; originally it was released by a handle in the centre, lifted inwards and jettisoned, but when Mod. 1336 is incorporated the hatch is enlarged and is opened by a handle at the port side. It opens inwards and is secured by a clip which holds the hatch up on the starboard side. It can also be opened from outside the aircraft.

66. **Crash exits**

(i) On Mks. 1, 3 and 10 aircraft, three push-out panels are fitted in the roof (one above the pilot, one just forward of the rear spar, and one forward of the mid-upper turret) except when Mod. 977 (which moves the mid-upper turret forward) is incorporated, in which case the third panel is deleted.

(ii) On Mk. 7 aircraft there are two push-out panels in the roof, one above the pilot, and one just forward of the rear spar.

67. **Dinghy**

(i) A dinghy stowed in the starboard wing may be released and inflated :—

    (a) from inside by pulling the release cord running along the fuselage roof aft of the rear spar.

    (b) from outside by pulling the loop on the starboard side, rear of the tail plane leading edge.

    (c) automatically by an immersion switch.

68. **Ditching**

If ditching is inevitable :—

    (a) The undercarriage should be kept retracted, but the flaps should be lowered 30° to reduce the touchdown speed as much as possible.

    (b) All crash exits should be opened (see para. 65).

    (c) Safety harnesses should be kept tightly adjusted and locked, but R/T plugs should be disconnected.

(d) If available the engines should be used to help make the touchdown in a tail-down attitude at as low a forward speed as possible.

(e) Ditching should be made along the swell, or into wind if the swell is not steep.

69. **Engine fire-extinguishers**

Each engine is provided with a fire-extinguisher system. When Mod. 1221 is incorporated four fire warning lights (one for each engine) on the front panel indicate if there is fire in an engine and the pilot is thus warned to stop the engine, feather the propeller and then press the appropriate fire-extinguisher button (22). When Mods. 1067 and 1314 are incorporated the fire warning lights are mounted on the respective feathering push-buttons (19), and if a fire warning light comes on, pressing the button also operates the fire-extinguisher system. The pilot should, however, press the fire-extinguisher button as well. If the warning light is not on, pressing the feathering pushbutton will not operate the extinguisher. The fire-extinguishers are also operated automatically by a crash switch.

70. **Emergency equipment**

(i) *Hand fire-extinguishers*

One on the starboard side of the air-bomber's compartment.

One on the port side of the pilot's seat.

One on the starboard side forward of the front spar.

One on the starboard side aft of the mid-upper turret.

One on the port side of the rear turret.

(ii) *Signal pistol*

This is stowed on top of the front spar ; the firing position is in the roof forward of the stowed position. The cartridges are stowed in spring clips on the starboard side of fuselage just forward of front spar.

(iii) *Crash axes*

One on port side of fuselage forward of main entrance door.

One on starboard wall in front of rear spar.

(iv) *First-aid equipment*

An outfit is stowed on the starboard side of the fuselage aft of the main door.

PART V

ILLUSTRATIONS and LOCATION OF CONTROLS

Location of controls not illustrated.

Service	*Location*
Undercarriage warning horn test pushbutton.	Behind pilot's seat.
Cross feed cock.	On floor, just forward of the front spar.
Priming pump and cock (if fitted).	In each inboard nacelle.
Air intake heat control.	Left of pilot's seat.
Radiator shutter switches.	On starboard cockpit wall.
Ground/flight switch.	On starboard side aft of front spar.
Cockpit heat controls.	One each side of the fuselage just forward of the front spar.
	Two adjustable louvres in fuselage nose.
Oxygen master valve.	At forward end of oxygen crate.
Camera pushbutton control.	On starboard rail of cockpit.
Reconnaissance flare stowage.	On either side of fuselage forward of flare chute.
Flame floats or sea markers stowage.	On either side of fuselage adjacent to flare chute.
Nitrogen system cock (if fitted).	On starboard side aft of front spar.
Flying controls locking gear.	Stowed on starboard side aft of front spar.
Dinghy release.	Fuselage roof aft rear spar.
Engine fire warning lights (if fitted).	Front panel.

FUSES

Service	*Location*
Bomb gear fuses.	Inside junction box at forward end of bomb aimer's compartment.
Oil and radiator thermometer fuses.	Pilot's auxiliary fuse panel.
Radio fuses.	Navigation panel.
Mid-upper and underturret, call lights, and, on early aircraft, beam approach fuses.	Mid-turret position.
General services.	Main electrical control panel.